Tips for Your Ice Cream Maker

Tips for Your Ice Cream Maker

CHRISTINE McFADDEN

7 9 10 8 6

First published in Hardback in 2008 by Ebury Press, an imprint of Ebury Publishing

Published in Paperback in 2013 by Ebury Press

Ebury Publishing is a Random House Group company

The Random House Group Limited Reg. No. 954009

Addresses for companies within the Random House Group can be found at: www.randomhouse.co.uk

A CIP catalogue record for this book is available from the British Library

Penguin Random House is committed to a sustainable future for our business, our readers and our planet. This book is made from Forest Stewardship Council® certified paper.

Printed and bound in Great Britain by Clays Ltd, St Ives plc

ISBN 978 0 09 195572 4

Note: The Department of Health advises that eggs should not be consumed raw. This book contains dishes made with raw or lightly cooked eggs. It is prudent for vulnerable people such as pregnant and nursing mothers, invalids, the elderly, babies and young children to avoid uncooked or lightly cooked dishes made with eggs. Once prepared, these dishes should be kept refrigerated and used promptly.

This book includes dishes made with nuts and nut derivatives. It is advisable for customers with known allergic reactions to nuts and nut derivatives and those who may be potentially vulnerable to these allergies, such as pregnant and nursing mothers, invalids, the elderly, babies and children, to avoid dishes made with nuts and nut oils. It is also prudent to check the labels of pre-prepared ingredients for the possible inclusion of nut derivatives.

Contents

introduction

Having an ice cream maker means you can enjoy the luxury of home-made ice cream as a year-round treat. It allows you to create glorious concoctions that can be ready to serve in as little as 20 minutes from start to finish.

You can turn store-cupboard ingredients, such as chocolate, nuts and biscuits, into impressive desserts, and take advantage of the intense flavours of fruit in season. If you've got Seville oranges left over from marmalade-making, you can use them to make a deliciously tangy ice cream. If there's a glut of strawberries or other soft fruit, a batch of mouth-watering sorbet or ice cream can be on the table in very little time. If you're feeling adventurous, there's a whole world of stunning savoury ices and luxurious unusual desserts for impressing your dinner guests.

An ice cream maker also allows you to control the texture of ice cream without any need for the emulsifiers and stabilisers found in commercial ices. Depending on the recipe, you can eat ices straight from the machine at the soft-scoop stage, or leave them in the freezer for an hour or two to harden. The choice is yours.

Modern ice cream machines are really easy to use. Just plug it in, wait for the container to get cold and you're all set to go. There's no tedious hand-cranking, mixing up brine or tripping over wires trailing from the freezer to an electrical socket.

There are hours of fun to be had experimenting with different ingredients and flavours, and making up your own special recipes. Once you've got the hang of the basics and churned your first batch, you'll be amazed at how easy it is to create something uniquely delicious.

getting started

- *Choosing a Machine*
- *Additional Equipment*
- *Manufacturer's Instructions*

choosing a machine

Nowadays, there are two basic machines to choose between. One type has a removable canister that requires freezing for 12 hours before use; the other comes with a fixed container and a built-in freezer unit. Both have a motorised churner that converts the mixture to ice cream.

When choosing a machine bear in mind the following: how often you are likely to use it; its capacity; the overall size; how large the hole is for pouring in the mixture; how easy it is to dismantle and clean and how simple the locking mechanisms. You may also be guided by budget; consider a less-expensive model to start with and upgrade once you've assessed how serious your ice cream making habit has become!

machines with a pre-frozen canister

Least expensive are the machines with a removable canister. The canister wall is insulated with a coolant that is liquid at room temperature and solid when frozen – similar to an ice pack. Once sufficiently chilled, the canister is locked in position in the machine and a paddle (or dasher) is attached to the top. The chilled mixture is poured in and the motor switched on. Depending on make, the motor rotates either the paddle or the canister until the mixture reaches the correct temperature and consistency, causing the rotation to stop – usually about 20 minutes.

The canisters generally come in two sizes: 1.1 litre/2 pints/4½ cups, or 1.5 litre/2½ pints/6 cups. During churning and freezing the volume of the mixture increases, so the size of the canister must be larger than the amount of ice cream to allow for expansion.

• Pros

These machines are the least expensive.

They are worth having if you want to make ice cream only occasionally and in small quantities.

• Cons

You have to remember to freeze the canister in advance. This can be overcome by keeping it permanently in the freezer.

You'll need to freeze the canister again if you want to make a second batch. You can get round this by having a second pre-frozen canister at the ready.

Canisters take up valuable freezer space.

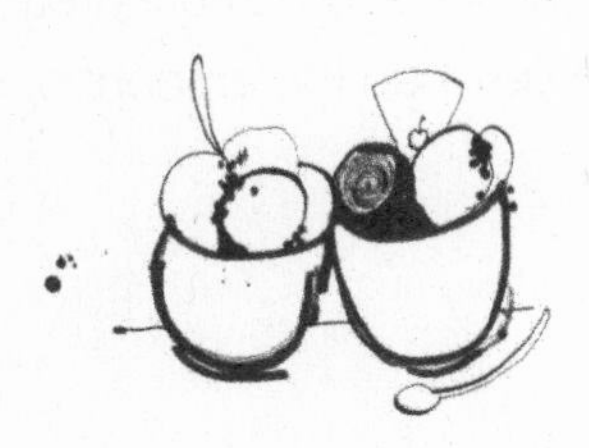

machines with a built-in freezer unit

With a sophisticated freezing mechanism, these machines are the connoisseur's choice of ice cream machine and this is reflected in the price.

They are quick to set up and easy to use. A compressor chills the container within five minutes of turning on the machine. The chilled mixture is poured in and a motorised paddle attached to the lid is plugged into the top of the machine. This paddle rotates and churns the mixture to the correct temperature and consistency, causing the rotation to stop. Depending on the recipe, this takes 15–40 minutes. You can then transfer the ice cream to a freezer-proof container, or eat straight from the container if you are going to eat it all in one go!

The motor has to rest for the same amount of time that it has been in action, then it's ready for the next batch.

Some models come with a smaller removable container that sits inside the fixed container. This requires a coolant of salt or alcohol to fill the gap between the containers, and the freezing process takes longer. However, it is easier to serve ice cream from a container that you can bring to the table. The removable container is also easier to clean.

- **Pros**

They are reliable.

The machine is ready within five minutes of turning on the power.

Batches can be made one after the other.

Some machines come with a special spatula and an extra container.

Top-quality commercial ice cream is expensive, so the machine may be a worthwhile investment, especially for families.

• Cons

The machine is much more expensive than machines with a pre-frozen canister. It's suitable only for the serious ice-cream maker.

The machine is too heavy to haul in and out of a cupboard, so it has to be kept permanently on the counter top where it takes up valuable space.

The machine must be kept level in order not to damage the freezing unit.

The fixed container is difficult to keep clean.

additional equipment

When making ice cream there are several items of kitchen equipment that will help in the process.

essential tools and utensils

The following checklist shows the tools and utensils that are used most often when making home-made ices. If your kitchen is reasonably well equipped you are likely to own most of them already.

Blender or liquidiser
Cling film
Electric hand-held mixer
Food processor
Fridge/freezer thermometer: for checking correct running temperature
Grater/zester: for fine wisps of citrus zest
Labels: freezer-proof, self-adhesive
Measuring spoons and cups

Measuring jugs: two each, large, medium
and small
Mixing bowls: large stainless steel or glass
Pen: freezer-proof marker
Prep bowls: several, small, for chopped ingredients
Saucepans: heavy-based
Scales: that are accurate to within 5 g
(scant ¼ oz)
Scoops: oval and round
Sieves: nylon, fine-meshed metal, small conical
Spatulas: plastic, rubber and heatproof silicone
Storage boxes: plastic
Sugar thermometer: for measuring temperature of
custard
Whisk: large balloon type for folding whipped cream
into custard
Wooden spoons

Quick Tip

Always use wooden or plastic utensils with your ice cream maker, never metal, to avoid damaging the canister.

optional extras

These tools and utensils are only really necessary if you want to launch into specialist ice cream making.

Bombe/ice cream moulds
Conical cone form: for making your own cones
Double saucepan: for foolproof egg-based sauces
Kulfi moulds/ice lolly moulds
Pizelle maker: for home-made cones and wafers
Saccharometer: for accurately measuring the ratio of sugar to liquid in sugar syrups

manufacturer's instructions

Unless you are an experienced cook, it's a good idea to follow the manufacturer's instructions to begin with. Most ice cream makers come with a free recipe book, and it's worth trying these out while you get to know your machine. You'll be able to achieve spectacular results even on your first attempt.

some simple science

- *What is Ice Cream?*
- *How does Ice Cream Freeze?*
- *What about Flavour?*

If you want to enjoy experimenting and developing your own recipes with confidence, it's well worth coming to grips with some simple chemistry and physics. Once you understand how these concepts relate to the freezing process, you'll be able to fine-tune ingredients, textures and flavours to create top-notch ice creams and sorbets that will have family and friends queuing up for more.

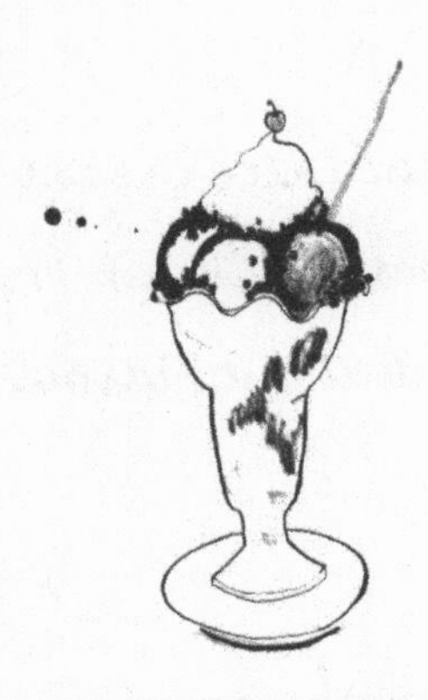

what is ice cream?

Ice cream is basically a frozen foam made up of tiny air pockets trapped between an intricate network of ice crystals, solid fat globules and a small amount of liquid containing dissolved milk solids, sugar and mineral salts. Strangely, this liquid never actually freezes.

Water ices, also called fruit ices or sorbets, have a similarly complex structure based on air, ice crystals and unfrozen liquid, but they contain fruit purée or juice, rather than milk solids and fat.

The key components of the frozen foam work together like a well-rehearsed orchestra, to contribute to the characteristics of the finished ice.

air

Although we can't see it, touch it or smell it, air is an indispensable ingredient. The tiny, trapped air pockets separate the solid and liquid parts of the mixture, resulting in a softer and lighter ice. Without it, the frozen mixture would be too dense to bite into.

Air affects the temperature, too. The more air an ice cream contains, the less cold it seems. This is because there are fewer ice crystals and less sugar solution to sustain the sensation of coldness. You may have noticed that cheap commercial ice cream tastes not quite cold enough and also feels light in the carton. This is because it has been pumped full of air, to increase the volume – and profits.

water

Like air, water is a vital ingredient; without it, there would be no ice cream. When the temperature of the mixture drops below freezing point (0°C/32°F), some of the water starts to freeze into solid ice crystals. These help stabilise the foam by trapping other components – fat, air and unfrozen liquid. The remaining water is needed as a carrier for dissolved milk solids, sugars, mineral salts and flavourings.

sugar

Sugar not only contributes to the flavour but also controls the texture and eating quality of ice cream. It does so by getting in the way of water molecules that want to clump together and form crystals, which they naturally start to do at freezing point (0°C/32°F). This, in turn, lowers the freezing temperature of the mixture, resulting in an ice cream that is nice and cold. The more sugar there is in proportion to water, the greater this effect will be. Sugar also stops ice crystals from growing too large and making the texture unpleasantly grainy and crystalline.

As the freezing point lowers, more and more of the water will eventually form ice crystals, but the sugar concentration in the small amount of remaining unfrozen liquid increases to the point where freezing becomes impossible. An ice made with intensely sweet undiluted elderflower syrup, for example, will remain a sweet sloppy mess, however long you leave it in the freezer, because it simply contains too high a concentration of sugar. The right concentration of sugar, however, gives an ice a lovely scoopable texture straight from the freezer. If there were no sugar, the ice cream would become a block of rock-hard ice.

Quick Tip

By experimenting with the ratio of sugar to water, you can alter the freezing point and therefore the texture of your mixture (see also Sugar Syrup, page 54).

fat

Fat from egg yolks, cream and milk provides richness, smoothness and body. It lubricates the mixture by forming globules that separate small ice crystals, preventing them from growing into large clumps. Ices with a lower fat content – those made with yogurt for example – therefore have a coarser texture than those made with high-fat cream and a rich egg custard. They also feel colder in the mouth because there are more ice crystals.

Water ices contain little, if any, fat, and so have a naturally coarse texture. This is smoothed by the amount of sugar syrup added and the churning and freezing process.

how does ice cream freeze?

As the temperature of the base mixture drops to freezing point or below, the molecules in the water gather together to form ice crystals. If left undisturbed, as in the early stages of still freezing (see Glossary, page 138), these crystals increase in size, producing a coarse-textured ice cream rather than a smooth one.

Rapid cooling and stir-freezing in an ice cream maker produces hundreds of tiny crystals that each have to share the available water molecules and therefore cannot increase in size. The churning action of the machine breaks up any clusters and ensures that the mixture freezes evenly. It also introduces air, which separates the crystals and stops them sticking together and increasing in size.

what about flavour?

The flavour of ice cream is part and parcel of our enjoyment. Without the right intensity of strawberry, chocolate or citrus fruit, for example, or the optimum balance of sweetness, tanginess or acidity, ice cream will still be good to eat, but it will not be superb.

Flavour is a combination of the smell of food and the basic tastes – sweet, acid, salty and bitter – that we experience through thousands of taste buds on the tongue, inside the cheeks and at the back of the mouth. When we chew food, the aromas and tastes are sent as chemical messages to the brain, which interprets them as flavour. The ability to smell is vital for experiencing flavour, as most of us who have had a stuffy cold will know. Once your sense of smell is reduced your food will have little flavour, although you may still be able to 'taste' some of its sweetness or acidity.

The problem with ice cream, and almost all chilled or frozen food, is that it has no smell. Think how good strawberries smell at room temperature, and how little aroma they have when chilled. Or try sniffing chocolate ice cream and lemon sorbet blindfold – there'll be no discernible difference.

The trick when making ices is to boost the flavour enough to compensate for the dulling effect of freezing. In practical terms this means that the flavour of a chilled mixture should pack more of a punch than you would normally expect. Once frozen, though, any excessive sweetness, acidity or other flavour will be tamed to the point where it tastes just right.

safe ice cream making

- *Cleaning*
- *Heating*
- *Cooling*

Food containing ingredients of animal origin, such as eggs, milk and cream, is particularly prone to contamination from harmful bacteria and other organisms which can cause anything from a stomach upset to serious illness. They thrive in wet and warm conditions and are merely held in check, but not destroyed, by freezing. It's essential, therefore, to follow good hygienic practice, particularly if you are serving ices to pregnant women, elderly people or infants.

cleaning

Before you begin, make sure that all the equipment – canister, paddle, lid – and the machine itself is scrupulously clean. Spatulas and serving scoops should also be cleaned properly. Manufacturers generally suggest warm soapy water for cleaning. Canisters that are prefrozen can't be put in a dishwasher, because of the coolant in the insulated walls.

machines with a fixed container

These are particularly difficult to clean since you cannot remove the container to wash it, nor can you fill it with water. (You would need to tilt the machine to pour away the water and this would damage the freezing mechanism.)

Though laborious, the best and safest way to clean a fixed container is to let the ice cream residue melt, then use plenty of dry paper towels to wipe away as much liquid as possible. Once the inside of the container looks fairly dry, dampen several wads of paper towel with hot water and a splash of detergent, and use these to wipe away the dried residue.

The next step is to use more wet paper towel to wipe up traces of detergent. Finally, wipe the container again with dry paper towel until the inside is completely dry and gleaming.

If you have made a mixture containing eggs, you might want to use an antibacterial wipe as well, but you will need to repeat the wet and dry paper towel process again to get rid of any residual taint from the wipe.

Quick Tip

For extra safety, clean everything immediately after making a batch of ice cream, and then repeat the process just before you use the machine again.

heating

When making ice cream containing eggs, there is a risk of infection from *salmonella*, a bacterium found in raw eggs that causes food poisoning. When making an egg-based ice cream make sure the custard is brought to at least 85°C/185°F, the temperature at which the bacterium is destroyed.

Sugar syrups should also be brought to boiling point to kill off any harmful organisms.

cooling

Once the custard is made, it must be cooled and refrigerated as quickly as possible before any harmful bacteria can proliferate. Speed the process by pouring the mixture from the hot saucepan into a cold metal bowl. Place the bowl in a larger one – a washing-up bowl is perfect – filled with ice cubes or very cold water. Regular stirring will also help bring down the temperature.

Wait until the mixture is completely cold before putting it in the fridge. A warm or even tepid mixture will cause the fridge temperature to rise, increasing the risk of food spoilage or unwanted bacteria.

Though not as prone to contamination, sugar syrups should be cooled quickly too, before storing in the fridge.

fridge temperature

The coldest part of your fridge should be between 0–5°C/ 32–41°F. If you do not know the temperature or the location of the coldest part, it's worth buying a fridge thermometer so you can monitor the fluctuations – you'll be surprised how the temperature changes. Keep the thermometer in the coldest part and check it regularly.

Never overload the fridge as this can block the circulation of cold air. And no matter how tempted, never put a not-quite-cold mixture in the fridge as you risk raising the temperature to unsafe levels.

chilling

Pour the cold mixture into a container and cover tightly with a lid or cling film to prevent contamination or odours from other items in the fridge. Leave to chill in the coldest part of the fridge for at least two hours.

Quick Tip

It's best not to leave ices containing cream or eggs to soften at room temperature. It's safer to soften them in the fridge for 20–30 minutes before serving, even though this takes a little longer.

enjoying your ices

- *Storing*
- *Tips for Storing*
- *Serving in Style*

storing

To maintain the mouth-watering texture, ices should be stored at a steady temperature between -18°C and -23°C (0°F and 10°F). Even the slightest rise in temperature will cause some ice crystals to melt. When the temperature drops again, the defrosted liquid finds its way to frozen crystals and merges with them, producing larger crystals as the liquid refreezes. Result: a coarse, grainy texture.

It's better to store ices in a proper freezer rather than the freezer unit in a fridge. The temperature of the freezer is lower and, since the door isn't opened as often as the fridge door, the temperature does not fluctuate so much.

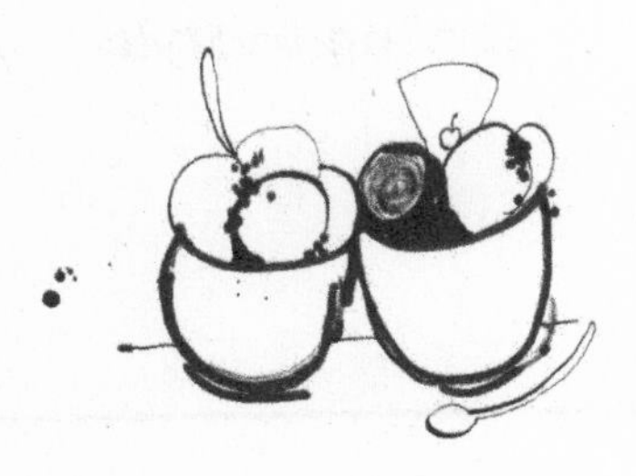

tips for storing

- Use shallow freezer-proof containers with tightly sealed lids.
- Leave at least 1 cm/½ in head space in the container to allow for expansion during freezing.
- Before putting on the lid, cover the surface of the ice with cling film, pressing down firmly to get rid of air pockets. This will protect the ice from free moisture settling on the surface and forming large ice crystals during storage.
- Label the container with the date. Most home-made ices should be eaten within a week or two while the flavour and texture are at their best. The more often you take them from the freezer to soften up for serving, the more the texture will coarsen during storage. It's better to make an amount that can be used up quickly.
- Ices containing raw egg white should be eaten within 48 hours.

serving in style

ideal temperature

There is something so pleasurable about scooping ice cream that is still cold and firm enough to resist pressure, but at exactly the right temperature for serving – that is, just before it reaches melting point. The softly rounded shape and the slightly roughened texture get the mouth watering in anticipation of the treat to come.

Many ice creams can be served directly after churning, but first should be left in the container to rest for 10 minutes with the machine turned off. Ices that have been hardened in the freezer, or stored for a length of time, should be put in the fridge to soften for 20–30 minutes before serving.

Quick Tip

It's a good idea to chill serving dishes or plates in advance. Ices will stay in peak condition for longer so you can enjoy every spoonful before it melts. Chilled glass and metal dishes look particularly refreshing covered with frosting as they come out of the fridge.

scoops

The best and easiest way to serve ices is with a professional metal scoop. Cheapest is the heavy aluminium oval scoop that is filled with a salt-based defrosting fluid to stop the ice sticking. The rim of the bowl is tapered so it easily penetrates hard-packed ices. These scoops are not dishwasher-proof so must be kept scrupulously clean.

The other type is a rounded mechanical scoop with a hemispherical bowl and a double handle or trigger mechanism that you squeeze to release the ice. Buy the best you can afford; cheap ones won't perform well.

shapes

- *The ball or 'boule'*: balls look good on top of sundaes or in cones; a multicoloured trio is particularly spectacular. A rounded mechanical scoop will make perfect balls that release without sticking.
- *The curl*: a thick curled ribbon of ice cream looks lovely with Chasummer fruits, and makes a change from the traditional ball. Dip an oval metal scoop in very hot water, let it heat for a moment, then quickly dry it with paper towel. Drag the edge of the hot scoop over the surface of the ice in one long movement towards you, forming the scooped ice into a curled roll as you drag, rather like using a butter curler.
- *The quenelle*: an oval shape that is a favourite with chefs, the quenelle sits elegantly on a plate, either fanned out in a star-shaped group, or singly to the side of other components of the dish. Dip two dessertspoons or tablespoons in cold water first, then use one to scoop up a mound of ice cream, and the other to fashion it into an oval. Turn the ice cream repeatedly between the spoons until you have the correct shape. Use the tip of the second spoon to push the oval gently on to a chilled plate.

tableware

It's fun to serve ices in traditional goblets and sundae dishes. Traditional metal dishes – the kind found in old-fashioned restaurants or hotel dining rooms – have a pleasantly retro feel when filled with plump balls of classic vanilla or chocolate ice cream. Tall glass goblets are good for displaying melon-sized balls of different-coloured sorbets. Long-handled sundae spoons are a must for completing the experience.

For a more modern presentation, small ovals of different-coloured ices look beautiful on a plain white plate, perhaps with some thinly sliced fruit or a gleaming sauce of contrasting colour.

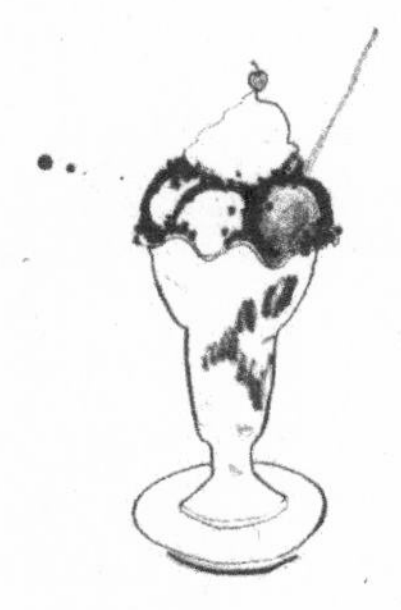

basic ingredients

- *Cream*
- *Milk*
- *Yogurt and Ricotta*
- *Eggs*
- *Sugar*
- *Fruit*
- *Alcohol*
- *Flavourings*

Ices fall essentially into two categories – ice creams based on dairy products, and water ices, also called sorbets, made with sugar syrup and flavourings. The dairy-based group often include a vanilla-flavoured custard made with milk, eggs and sugar. Water ices invariably include lemon juice.

cream

Whipping cream (36% fat) gives a smooth texture and a pleasant creamy flavour. The best choice for custard-based ices.
Double cream (48% fat) provides a very rich flavour but can taste slightly greasy and sometimes masks other flavours. It is easy to over-churn double cream, so you may end up irrevocably with frozen butter.

Depending on the recipe, cream can be heated with milk and eggs at the custard-making stage, or, for added volume and a slightly lighter texture, it can be lightly whipped and folded into the custard once this is cooled and chilled.

milk

Full cream milk (3.8% fat) gives a full-bodied flavour, as does evaporated milk (9% fat).
Skimmed and semi-skimmed milk contain hardly any fat and therefore do not produce good ice cream.

yogurt and ricotta

Full-fat strained Greek yogurt (10% fat) is a lower-fat alternative to cream. It is convenient to use and creates a pleasantly creamy ice despite containing less fat.

Low-fat yogurt has a thin, watery flavour and little body when frozen.

Ricotta cheese (10% fat) is another option for the calorie conscious. The slightly granular texture is improved by mixing with a spoonful or two of cream.

eggs

Egg yolks emulsify fats and water, providing ice cream with extra richness and a smooth texture. For yolks with the best flavour, it is worth using organic or traditional free-range eggs.

Lightly beaten raw egg white used to be routinely added to sorbets to stabilise and lighten the mixture. However, unless the mixture is very dense and freezes to a rock-hard consistency, egg white is not really necessary. Leaving it out also neatly avoids the risk of salmonella poisoning (see Safe Ice Cream Making, page 29).

Use left-over egg whites to make small meringues either to serve as an accompaniment to ice cream or to crumble into the mixture during churning. Egg whites can also be used to make crisp little biscuits such as tuiles.

sugar

Sugar provides sweetness but also creates texture in a frozen ice (see Some Simple Science, page 19).

Granulated sugar has medium-sized crystals and is fine for mixtures that are heated, since heat will help dissolve the crystals.

Caster sugar or powdery icing sugar are better for unheated mixtures, since they dissolve without any heat.

fruit

Since chilling and freezing dulls flavours, the best results come from top-quality fruit that is perfectly ripe and actually smells of fruit. A splash of lemon juice will enhance the flavour once the fruit is chopped or puréed. Different fruits have varying levels of sweetness and this will affect the amount of syrup, sugar and/or water to be added.

alcohol

Alcohol never freezes, as those who keep a bottle of vodka at the ready in the freezer will know. Like sugar, alcohol lowers the freezing point of ices, but, if you add too much, the ice will never freeze. As a rule of thumb, alcohol should be no more than about 10 per cent of the total mixture. Spirits tend to lower the freezing point the most, and wines the least. Fortified wines such as sherry and port come somewhere in between.

The following amounts can be added to 1 litre/1¾ pints/4 cups of ice cream mixture before churning and freezing:

Up to 3 tablespoons of spirits
Up to 6 tablespoons of fortified wines
Up to 125 ml/4 fl oz/½ cup of wine or champagne

flavourings

Vanilla is routinely added to most custard-based ice creams. Use a vanilla pod for the best flavour.

Lemon juice is often used in syrup-based ices to accentuate and brighten the flavour.

Salt is surprisingly useful in some ice cream recipes. A couple of pinches are undetectable but will round out and enhance the overall flavour.

After that, the choice is endless. Chocolate, coffee, nuts, dried fruit, crumbled biscuits, liqueurs, herbs and spices all create unique flavours for you to experiment with.

Quick Tip

Ideally, any additional solid ingredients should be chilled or pre-frozen to prevent them from sinking to the bottom of the mixture. Add them once the mixture has started to thicken.

basic recipes

- *Vanilla Custard*
- *Sugar Syrup Concentration*
- *Sugar Syrup*

vanilla custard

If you plan to use your ice cream maker regularly, it is a good idea to make up and chill a batch of custard in advance so that you're all set to go when the mood takes you. For the best flavour, use a vanilla pod and its sticky black seeds rather than vanilla extract. Allow plenty of time for cooling and chilling the custard. It can be kept in the fridge for 2–3 days.

Makes about 350 ml/12 fl oz/1½ cups

300 ml/½ pint/1¼ cups whole milk
100 g/3½ oz/½ cup sugar
½ vanilla pod
3 egg yolks, preferably organic

Put the milk in a saucepan with about two-thirds of the sugar. Split the vanilla pod lengthways with the tip of a knife. Scrape out the seeds and add these to the pan along with the pod. Stir over medium heat for about 5 minutes until steaming

(80°C/175°F), but do not let it boil. Remove from the heat and leave to infuse.

Beat the egg yolks with the remaining sugar for 4–5 minutes until very pale and creamy. Fish the vanilla pod out of the warm milk. Gradually whisk the milk into the thickened yolks. Pour the custard back into the pan and stir over medium–low heat for about 5 minutes until thickened (85°C/185°F). Take care not to let the mixture boil, or you will end up with scrambled eggs.

Cool the custard quickly by pouring into a bowl immersed in a larger bowl of ice cubes or very cold water. Once cold, cover the surface with cling film to prevent a skin forming. Ideally, chill in the fridge for at least 4 hours or up to 24 hours before churning to allow the flavour to develop.

sugar syrup concentration

Many recipes for sugar syrup give different ratios of water to sugar. They also instruct you to boil the syrup for varying lengths of time, which serves only to evaporate some of the water and concentrate the syrup.

It is simpler and easier to use an equal ratio of water to sugar, as in the recipe below, rather than mixing up different proportions for different ices. If you need a less-concentrated syrup, just add more water to the rest of the ingredients when making the basic mixture.

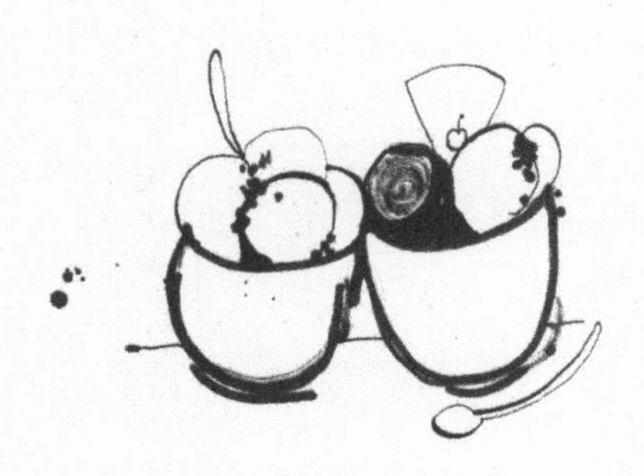

ratios for smaller quantities

If you want a smaller quantity of syrup, use the following amounts of sugar and water:

Amount of syrup required	Water	Sugar
700 ml/1¼ pints/2¾ cups	400 ml/14 fl oz/1¾ cups	400 g/14 oz/2 cups
600 ml/1 pint/2½ cups	350 ml/12 fl oz/scant 1½ cups	350 g/12 oz/1¾ cups
500 ml/18 fl oz/2 cups	300 ml/½ pint/1¼ cups	300 g/10½ oz/1½ cups
400 ml/14 fl oz/1¾ cups	225 ml/8 fl oz/scant 1 cup	225 g/8 oz/rounded 1 cup
250 ml/9 fl oz/1 cup	150 ml/¼ pint/⅔ cup	150 g/5½ oz/¾ cup
200 ml/7 fl oz/scant 1 cup	125 ml/4 fl oz/½ cup	125 g/4½ oz/rounded ½ cup

sugar syrup

If you plan to use your ice cream maker regularly, it's a good idea to make up and chill a quantity of syrup in advance so that you're all set to go when the mood takes you. It can be kept in a screw-top jar in the fridge for up to two weeks. The syrup needs only to be brought briefly to the boil to destroy any harmful organisms, and then it's ready.

Makes about 850 ml/1½ pints/3½ cups

500 ml/18 fl oz/2 cups water
500 g/1 lb 2 oz/2½ cups sugar

Pour the water and sugar into a saucepan. Stir over medium heat until the sugar has dissolved and the solution is clear. Remove from the heat and leave until completely cold. Pour into a clean screw-top jar and store in the fridge.

Quick Tips

Before you begin churning and freezing make sure the mixture is thoroughly chilled. This reduces the churning time and means there is less chance of cream-based mixes turning into butter.

Make sure the removable canister from a machine without a built-in freezer unit has been in the freezer for at least 12 hours.

A machine with a built-in freezer unit should be turned on 5 minutes before you start to churn. This will chill the container and speed up thickening and freezing.

making cream ices

- *Hints for Making Cream Ices*
- *Cream Ices Recipes*

This chapter includes favourite custard-based classics such as Vanilla Ice Cream and Raspberry Ripple, as well as some more unusual flavours. There are also very simple ices based on cream without custard. They are very quick and easy to make – perfect if you suddenly feel a pressing need for some home-made ice cream.

Most of the ice creams can be eaten straight from the machine directly after churning and freezing. If you prefer a harder texture, just put them in the freezer for a few hours. They should be put in the fridge to soften about 30 minutes before you want to serve them.

hints for making cream ices

- It's fine to use a standard carton of cream (284 ml) if you don't have exactly 300 ml/½ pint/1¼ cups.
- When whipping cream, take care not to over-beat or it will become hard and buttery when churned.
- If using fruit purée, bear in mind that the purée should be thick and intensely flavoured so as not to dilute the cream or custard.

vanilla ice cream

This classic ice cream forms the basis for many others. Make sure the custard is thoroughly chilled before you add the cream.

Makes about 950 ml/1¾ pints/3¾ cups

300 ml/½ pint/1¼ cups whipping cream
1 quantity chilled Vanilla Custard (page 56)

Lightly whip the cream until the beaters leave a faint trail when lifted from the cream. Stir the cream into the vanilla custard, mixing well. Churn and freeze in an ice cream maker. Once thickened, either serve right away, or store in the freezer to harden.

chocolate ice cream

This is a simple-to-make everyday chocolate ice cream. Perfect for midnight snacking.

Makes about 1 litre/1¾ pints/4 cups

4 tbsp unsweetened cocoa powder
115 g/4 oz/rounded ½ cup sugar
seeds from a split vanilla pod
350 ml/12 fl oz/scant 1½ cups whole milk
250 ml/9 fl oz/1 cup whipping cream

Put the cocoa, sugar, vanilla seeds and a little of the milk in a small bowl and mix to a thin paste. Pour the rest of the milk into a saucepan and stir in the paste. Keep stirring briskly, pressing out clumps of powder with the back of a wooden spoon, until the cocoa is smoothly incorporated into the milk. Bring to the boil, then reduce the heat and simmer gently for 5 minutes, stirring all the time. Pour the mixture into a bowl, then place the bowl in a larger bowl of ice cubes or very cold water so that the mixture cools quickly. Lightly whip the cream until the beaters leave a faint trail when lifted from the cream. Stir the cream into the cold cocoa mixture. Cover and chill for at least 2 hours. Churn and freeze in an ice cream maker. Once thickened, either serve right away, or store in the freezer to harden.

raspberry ripple

A wonderfully flamboyant ice cream – classic vanilla streaked with ripples of tangy fresh raspberry purée.

Makes about 1 litre/1¾ pints/4 cups

250 g/9 oz raspberries
juice of ½ lemon
3 tbsp icing sugar
300 ml/½ pint/1¼ cups whipping cream
1 quantity cold Vanilla Custard (page 56)

Purée the raspberries in a food processor, then push through a nylon sieve to remove the seeds. Stir in the lemon juice and sugar. Cover and chill for 1 hour. Meanwhile, lightly whip the cream until the beaters leave a faint trail when lifted from the cream. Add the custard, mixing well. Cover and chill. Churn the cream mixture in an ice cream maker for 15–20 minutes, until almost frozen. While still churning, pour in the raspberry purée. Churn briefly until the cream is streaked with ripples of pink. Serve right away, or store in the freezer to harden.

strawberry ice cream

Diced strawberries add delicious nuggets of flavour and texture to this classic ice cream. Use really ripe strawberries that actually smell of strawberries.

Makes about 1 litre/1¾ pints/4 cups

450 g/1 lb ripe strawberries
100 g/3½ oz/½ cup caster sugar
juice of ½ lemon
300 ml/½ pint/1½ cups whipping cream

Briefly dunk the strawberries in water with their hulls still attached. Drain and dry on paper towel, then remove the hulls. Slice the strawberries in half lengthways and put in a shallow bowl. Sprinkle with a tablespoon of the sugar, and the lemon juice. Leave to marinate at room temperature for 30 minutes to allow the flavour to develop. Set aside about one-third of the strawberries and chop these into small dice. Purée the rest in a food processor with the remaining sugar. Lightly whip the cream until the beaters leave a faint trail when lifted from the cream. Add the purée and diced strawberries, stirring until evenly mixed. Cover and chill for at least 2 hours. Churn and freeze in an ice cream maker. Once thickened, either serve right away, or store in the freezer to harden.

• variation

Add 1 tablespoon crushed black peppercorns to the sugar and lemon juice marinade. Strawberries and black pepper go together remarkably well.

banana ice cream

This is a favourite with kids – easy to make and quick to freeze. Be sure to use very ripe sweet bananas.

Makes about 600 ml/1 pint/2½ cups

3–4 ripe bananas, preferably organic, about 500 g/1 lb 2 oz in total
5 tbsp sugar
juice of ½ lemon
juice of 1 small orange
175 g/6 oz can evaporated milk

Slice the bananas and put in a food processor with the sugar and lemon and orange juices. Purée until smooth then tip into a bowl. Stir in the evaporated milk. Cover and chill for at least 2 hours. Churn and freeze in an ice cream maker. Once thickened, either serve right away, or store in the freezer to harden.

coconut and stem ginger ice cream

Coconut milk makes a very dense cold ice cream that is delicious after a spicy meal.

Makes about 850 ml/1½ pints/3½ cups

300 ml/½ pint/1½ cups whipping cream
400 g/14 oz can coconut milk
6 pieces stem ginger, finely chopped
8 tbsp syrup from the stem ginger jar
juice of 1 lime

Lightly whip the cream until the beaters leave a faint trail when lifted from the cream. Add the coconut milk and mix well. Stir in the chopped stem ginger, syrup and lime juice. Cover and chill for at least 2 hours. Churn and freeze in an ice cream maker. Once thickened or store in the freezer to harden.

mango kulfi

This is an Indian ice cream made by lengthy boiling of milk until reduced to a rich, thick cream. You will need a ripe mango with a good sweet flavour. Failing this, use 300 g/10½ oz/1¼ cups canned mango purée.

Makes about 850 ml/1½ pints/3½ cups

1.5 litres/2¾ pints/6 cups whole milk
1 tsp arrowroot or rice flour
1 large ripe mango, weighing about 450 g/1 lb
juice of 1 lime
5 tbsp sugar

Bring the milk to the boil in a heavy-based saucepan, then simmer, stirring often, for 1–2 hours, until reduced by half. Mix the arrowroot to a thin, smooth paste with some of the milk. Stir the paste into the milk. Simmer over low heat, stirring constantly and without allowing the mixture to boil, until thickened to a consistency of pouring cream. Pour into a bowl and leave to cool. Meanwhile, peel the mango and chop the flesh. Put in a food processor with the lime juice and sugar, and purée until smooth. Mix with the cold thickened milk. Cover and chill for at least 2 hours. Churn and freeze in an ice cream maker. Once thickened, either serve right away, or store in the freezer to harden.

quick pistachio kulfi

This is a simpler version of the traditional kulfi – an irresistibly rich golden-brown ice cream, almost like frozen toffee.

Makes about 850 ml/1½ pints/3½ cups

400 g/14 oz can condensed milk
2 x 175 g/6 oz cans evaporated milk
350 ml/12 fl oz/scant 1½ cups whipping cream
seeds from 6–8 cardamom pods, crushed
85 g/3 oz/½ cup shelled pistachio nuts
pinch of salt

Slowly bring the canned milks and whipping cream to the boil in a heavy-based saucepan, stirring constantly to prevent sticking. Add the cardamom seeds, then simmer for 20 minutes over low heat, stirring constantly, until reduced by about one-third. Pour into a bowl and leave to cool. Chop the pistachio nuts fairly finely, then put in a blender and whiz to a powder. Stir into the cold milk along with a pinch of salt. Cover and chill for at least 2 hours. Churn and freeze in an ice cream maker. Once thickened, either serve right away, or store in the freezer to harden.

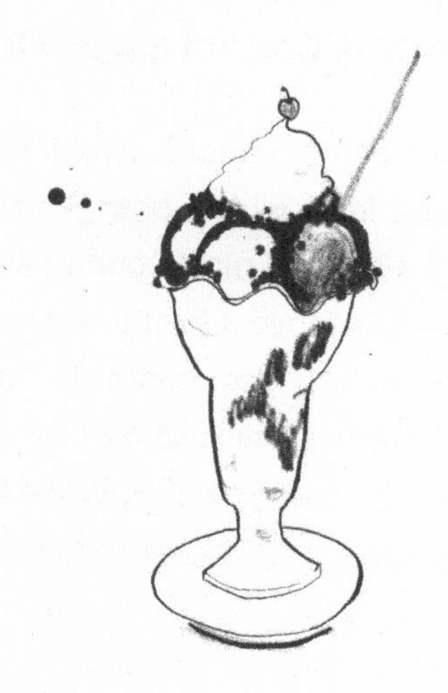

seville orange ice cream

Serve this palate-tingling ice with a scoop of Chocolate Sorbet (page 79). The contrasting flavours and textures are superb.

Makes about 1 litre/1¾ pints/4 cups

5–6 Seville oranges
250 ml/9 fl oz/1 cup whipping cream
1 quantity cold Vanilla Custard (page 50)

Grate the zest from two of the oranges, avoiding any bitter white pith. Squeeze the juice from all the oranges and strain into a bowl. You need about 150 ml/¼ pint/⅔ cup of juice. Lightly whip the cream until the beaters leave a faint trail when lifted from the cream. Stir the cream into the custard, mixing well. Add the orange juice and zest. Cover and chill for at least 2 hours. Churn and freeze in an ice cream maker. Once thickened, either serve right away, or store in the freezer to harden.

christmas ice cream

Packed with nuts and plump dried fruit, this deliciously creamy ice is an excellent alternative to Christmas pudding. Start the recipe the day before to allow time for soaking the fruit.

Makes about 1 litre/1¾ pints/4 cups

25 g/1 oz/¼ cup dried cranberries
25 g/1 oz/¼ cup mixed raisins and sultanas
55 g/2 oz/½ cup finely chopped mixed peel
100 ml/3½ fl oz/scant ½ cup marsala, rum or brandy
200 ml/7 fl oz/scant 1 cup whipping cream
250 ml/9 fl oz/1 cup cold Vanilla Custard (page 50)
40 g/1½ oz/¼ cup macadamia nuts, chopped
40 g/1½ oz amaretti biscuits, crumbled

Put the cranberries, raisins, sultanas and mixed peel in a bowl. Add the marsala and leave to soak for 24 hours. Lightly whip the cream until the beaters leave a faint trail when lifted from the cream. Drain the fruits and stir into the cream, along with the custard. Cover and chill for at least 2 hours. Add the nuts and crushed amaretti biscuits. Churn and freeze in an ice cream maker. Once thickened, either serve right away, or store in the freezer to harden.

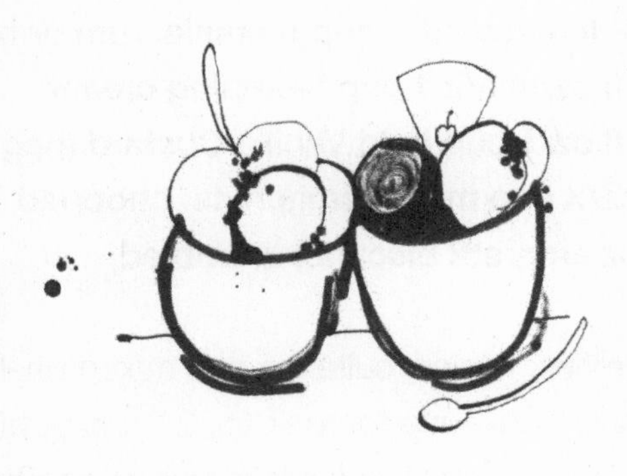

making water ices

- *Hints for Making Water Ices*
- *Water Ices Recipes*

hints for making water ices

- Water ices, or sorbets, are excellent made with intensely flavoured acidic fruits such as citrus fruits, passion fruit, berries and currants.
- The flavour is enhanced by a squeeze of lemon juice.
- Water ices based on fruit juice rather than purée need especially thorough churning to prevent ice crystals forming.
- Raw egg white is sometimes used in sorbets to lighten the texture or stabilise the consistency. However, it is not really necessary with an ice cream maker which does the job perfectly well.

lemon sorbet

Refreshing and cleansing, this classic sorbet is easy to make and always popular.

Makes about 750 ml/1¼ pints/3 cups

3–4 lemons, preferably unwaxed
400 ml/14 fl oz/1¾ cups Sugar Syrup (page 54)
250 ml/9 fl oz/1 cup water

Finely grate the zest from one of the lemons, taking care not to include any of the bitter white pith. Put the zest and sugar syrup into a saucepan and bring to the boil. Remove from the heat and leave to infuse until completely cold. Squeeze the juice from the lemons and strain through a fine-meshed sieve. Stir the strained juice and the water into the cold syrup. Cover and chill for at least 2 hours. Pour through a sieve to remove the grated zest. Churn and freeze in an ice cream maker. Once thickened, store in the freezer to harden.

• variations

lime sorbet: zest of 2 limes, 400 ml/14 fl oz/1¾ cups Sugar Syrup, juice of 5–6 limes (150 ml/¼ pint/⅔ cup), juice of ½ lemon, 250 ml/9 fl oz/1 cup water.
Makes about 750 ml/1¼ pints/3 cups

orange sorbet: zest of 1 orange, 400 ml/14 fl oz/1¾ cups Sugar Syrup, juice of 4–5 medium oranges (400 ml/14 fl oz/1¾ cups), juice of 1 lemon, 150 ml/¼ pint/⅔ cup water.
Makes about 850 ml/1½ pints/3½ cups

clementine sorbet: zest of 2 clementines, 400 ml/14 fl oz/ 1¾ cups Sugar Syrup, juices of 14 clementines, 1 orange, 1 lemon (400 ml/14 fl oz/1¾ cups total), 150 ml/¼ pint/⅔ cup water.
Makes about 850 ml/1½ pints/3½ cups

chocolate sorbet

Less creamy than chocolate ice cream, chocolate sorbet is still devastatingly rich and very cold.

Makes about 750 ml/1¼ pints/3 cups

60 g/2¼ oz/½ cup unsweetened cocoa powder
seeds from 1 vanilla pod, split
150 g/5½ oz/½ cup caster sugar
pinch of salt
600 ml/1 pint/2½ cups water
85 g/3 oz plain chocolate (at least 75% cocoa solids), broken into small pieces

Put the cocoa powder, vanilla seeds, sugar and salt in a bowl. Add a spoonful or two of the water and mix to a thin paste. Pour the remaining water into a saucepan and add the paste. Bring to the boil, stirring constantly. Reduce the heat to a very gentle simmer, add the chocolate and simmer for about 4 minutes, stirring constantly. Remove from the heat and pour into a bowl. Leave to cool, then cover and chill for at least 2 hours. Churn and freeze in an ice cream maker. Once thickened, store in the freezer to harden.

raspberry and elderflower sorbet

Raspberries make a superb, intensely flavoured sorbet. The elderflower adds a hint of floweriness.

Makes about 750 ml/1¼ pints/3 cups

400 g/14 oz raspberries
3 tbsp lemon juice
115 g/4 oz/scant 1 cup icing sugar
3 tbsp elderflower cordial
5 tbsp water

Purée the raspberries in a food processor, then push the pulp through a fine nylon sieve to remove the seeds. Stir in the lemon juice, icing sugar, elderflower cordial and water, mixing well. Cover and chill for at least 2 hours. Churn and freeze in an ice cream maker. Once thickened, store in the freezer to harden.

blueberry and honey sorbet

This works best made with a mild-flavoured honey, such as acacia or orange blossom.

Makes about 750 ml/1¼ pints/3 cups

200 ml/7 fl oz/scant 1 cup orange juice (from 3 medium oranges)
4 tbsp lemon juice
5 tbsp clear honey
500 g/1 lb 2 oz blueberries

Strain the orange and lemon juices, and mix with the honey, stirring until the honey is completely dissolved. If necessary, warm gently but do not let the mixture boil. Put the blueberries in a food processor with the honey solution and purée until smooth. Push through a nylon sieve to remove the seeds. Cover and chill for at least 2 hours. Churn and freeze in an ice cream maker. Once thickened, store in the freezer to harden.

apple sorbet

This sorbet is a pretty pale green, but it's important to work quickly to prevent the apples from browning.

Makes 750 ml/1¼ pints/3 cups

3 Bramley apples, weighing about 800 g/1 lb 12 oz in total
juice of 2 lemons
250 ml/9 fl oz/1 cup water
175 g/6 oz/scant 1 cup caster sugar

Quarter and core the apples but do not peel them. Chop the flesh into small cubes, tossing them in the lemon juice and water as you work. Immediately tip the mixture into a food processor along with the sugar. Purée for a few minutes until very smooth. Tip the purée into a bowl, cover with cling film pressed over the surface and chill for 30 minutes. Check the flavour and add more sugar or lemon juice if necessary. Churn and freeze in an ice cream maker. Once thickened, store in the freezer to harden.

• variation

Add a few leaves of chopped fresh mint when you start to churn and freeze.

three-melon sorbet

Enjoy this sorbet within a day of making it, while the flavour is at its best. Make sure the melons are perfectly ripe – they should smell and taste strongly of melon. Fruit juice concentrate can be found in health-food stores.

Makes about 1 litre/1¾ pints/4 cups

700 g/1½ lb wedge of watermelon
juice of 2 limes
75 ml/2½ fl oz/⅓ cup exotic fruit juice concentrate
1 large green-fleshed melon, such as Ogen or Galia
1 large orange-fleshed melon, such as Charentais or Canteloupe

Cut two thin segments from the watermelon, lengthways, leaving the rind attached. Cover with cling film and put in the fridge. Mix the lime juice and exotic fruit juice concentrate. Remove the seeds and rind from the remaining chunk of watermelon. Slice the other two melons in half round their middles and scoop out the seeds. Cut the flesh into segments and remove the rind. Keeping the three colours separate, chop the flesh of all three melons into cubes. Still keeping each colour separate, tip each batch into a blender with a third of the fruit juice solution, and purée until very smooth. Churn and freeze in three separate batches in an ice cream maker. Once frozen, store in the freezer for no more than 24 hours. When ready to serve, slice the reserved watermelon into triangles leaving the rind attached. Spoon scoops of each colour sorbet into tall chilled glasses and decorate with a watermelon wedge.

pomegranate sorbet

A stunning garnet-red sorbet with a smooth, velvety texture. It is delicious served with fresh figs or a compote of juicy pears. Alternatively, serve it as a savoury starter with chilled chopped cucumber and mint.

Makes about 600 ml/1 pint/2½ cups

6 large red-skinned pomegranates
125 g/4½ oz/rounded ½ cup caster sugar
3 tbsp lemon juice

Slice the pomegranates in half horizontally. Use the citrus attachment of an electric juicer, or a citrus press, to extract the juice in the same way that you would an orange. You will need 500 ml/18 fl oz/2 cups of juice. Stir the sugar into the juice until it dissolves. Add the lemon juice, then cover and chill for at least 2 hours. Churn and freeze in an ice cream maker. Once thickened, store in the freezer to harden.

Quick Tip

To extract maximum juice, roll the pomegranates on the work top before slicing them. Press firmly to soften the seeds and release the juice.

If you don't have fresh pomegranates, use a carton of pomegranate juice, but make sure it is unsweetened.

passion fruit sorbet

Passion fruit are expensive but it's hard to beat their heady flavour and aroma. Push the boat out and buy the plumpest fruit you can find. Perfect for a special occasion.

Makes about 600 ml/1 pint/2½ cups

16 plump passion fruit
400 ml/14 fl oz/1¾ cups Sugar Syrup (page 54)
150 ml/¼ pint/⅔ cup water
juice of 4 limes

Slice the passion fruit in half and scoop out the pulp with a teaspoon. You should have about 200 ml/7 fl oz/scant 1 cup of pulp. Mix with the sugar syrup, water and lime juice. Cover and chill for at least 2 hours. Push the mixture through a nylon sieve, pressing hard with the back of a wooden spoon to extract as much juice as possible. Churn and freeze in an ice cream maker. Once thickened, store in the freezer to harden.

• variation

For a creamier ice, stir in 300 ml/½ pint/1¼ cups whipping cream, lightly whipped, after sieving the mixture.

making yogurt and ricotta ices

- *Hints for Making Yogurt and Ricotta Ices*
- *Yogurt and Ricotta Ices Recipes*

Making ice cream with thick strained Greek yogurt or soft creamy ricotta cheese is amazingly easy and convenient.

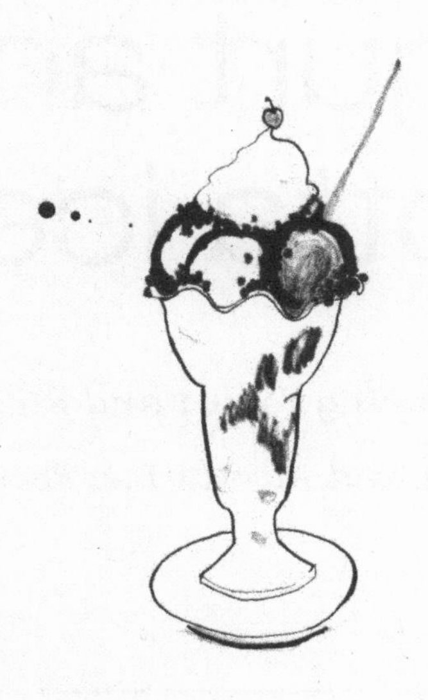

hints for making yogurt and ricotta ices

- Since yogurt and ricotta thicken the mixture beautifully there's no need to spend time making custard and waiting for it to cool and chill.
- You can even make great ice cream with a pot of yogurt or ricotta and some good-quality jam.
- Both contain 10% fat which provides delicious creaminess but with fewer calories than whipping cream.

cherry yogurt ice

A mixture of dried and fresh cherries gives this ice a really intense flavour.

Makes about 850 ml/1½ pints/3½ cups

550 g/1¼ lb cherries, pitted
75 g/1¾ oz/½ cup dried cherries
175 g/6 oz/scant 1 cup caster sugar
250 ml/9 fl oz/1 cup Greek strained yogurt

Put the fresh and dried cherries in a food processor. Add the sugar and purée until smooth. Tip into a bowl then fold in the yogurt. Chill for at least 2 hours. Churn and freeze in an ice cream maker. Once thickened, either serve right away, or store in the freezer to harden.

plum and raisin yogurt ice

Purple plums and plump raisins add mouth-watering texture to this rich, creamy ice.

Makes about 850 ml/1½ pints/3½ cups

50 g/1¾ oz/⅓ cup raisins
500 g/1 lb 2 oz dark-skinned plums
600 ml/1 pint/2½ cups water
200 g/7 oz/1 cup sugar
finely grated zest of 1 orange
1 tbsp lemon juice
250 ml/9 fl oz/1 cup Greek strained yogurt

Soak the raisins in warm water for 2–3 hours or until plump. Cut the plums in half round the indentation and through to the stone. Twist the two halves sharply in opposite directions to loosen the stone, then scoop it out with a knife. Slice the flesh lengthways into thin segments and put in a large bowl. Heat the water, sugar and orange zest in a saucepan, stirring until the sugar has dissolved. Raise the heat and boil for 7–10 minutes until the bubbles look syrupy. Immediately pour the syrup over the plums. Leave to cool. Strain the syrup and set aside. Finely chop about a quarter of the plums. Purée the rest in a food processor, along with the lemon juice. Tip into a large bowl and mix in the chopped plums. Drain the raisins and add them to the bowl. Fold in the yogurt and 250 ml/9 fl oz/1 cup of the reserved syrup. Chill for at least 2 hours. Churn and freeze in an ice cream maker. Once thickened, either serve right away, or store in the freezer to harden.

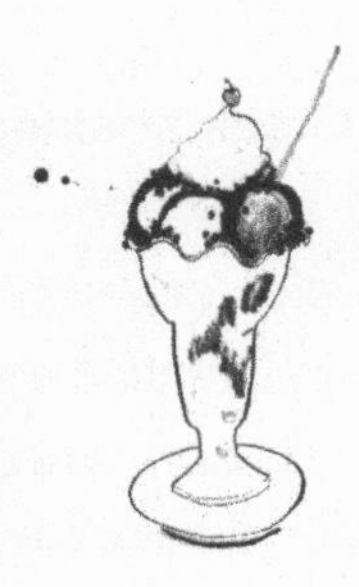

redcurrant yogurt ice

This vibrant pink creamy ice captures the essence of redcurrants. Eat within a day or two of making, while the flavour is at its best.

Makes about 850 ml/1½ pints/3½ cups

450 g/1 lb redcurrants, stalks removed
250 ml/9 fl oz/1 cup cold Sugar Syrup (page 54)
150 ml/¼ pint/⅔ cup whipping cream
150 ml/½ pint/⅔ cup Greek strained yogurt

Purée the redcurrants in a food processor or blender, then push through a fine nylon sieve and discard the debris. Fold the purée into the cooled syrup. Whip the cream until it just holds its shape. Stir the yogurt into the cream, followed by the redcurrant purée. Chill for at least 2 hours. Churn and freeze in an ice cream maker. Once thickened, either serve right away, or store in the freezer to harden.

apricot, pistachio and cardamom yogurt ice

Top-quality ripe apricots are a must for this cardamom-scented ice. The flavour of cardamom may seem quite strong at first, but it will be subdued once the ice is frozen.

Makes about 750 ml/1¼ pints/3 cups

500 g/1 lb 2 oz ripe apricots
500 ml/18 fl oz/2 cups water
350 g/12 oz/1¾ cups sugar
crushed seeds from 3 cardamom pods
juice of 1 lemon
50 g/1¾ oz/½ cup shelled pistachio nuts, chopped
150 ml/⅔ cup Greek yogurt, preferably organic

Slice the apricots in half round the indentation and scoop out the stones. Put in a saucepan with the water, 300 g/10½ oz/1½ cups of the sugar, and the crushed cardamom seeds. Bring to the boil, then reduce the heat and simmer for 10–15 minutes until very soft. Drain, discarding the liquid. Purée in a food processor, tip into a bowl and leave to cool. Stir in the lemon juice, yogurt and the pistachios. Taste, and add the remaining sugar if necessary. Cover and chill for at least 2 hours. Churn and freeze in an ice cream maker. Once thickened, either serve right away, or store in the freezer to harden.

chilli and guava ricotta ice

A schizophrenic taste sensation – very cold ice cream followed by a slow after-burn from the chilli. The Habanero chilli has a particularly pleasant fruity flavour, but any medium-sized fresh red chilli will do.

Makes about 1 litre/1¾ pints/4 cups

140 g/5 oz/⅔ cup sugar
300 ml/½ pint/1¼ cups water
1 Habanero chilli, deseeded
4 guavas, weighing about 800 g/1 lb 12 oz in total
juice of 2 limes
250 g/9 oz/1 cup ricotta cheese
6 tbsp whipping cream

Put the sugar, water and chilli in a small saucepan. Bring to the boil, stirring to dissolve the sugar, then simmer briskly for 5 minutes until the bubbles look syrupy. Leave to cool, then discard the chilli. Quarter the guavas, peel and cut into chunks. Purée in a food processor with the lime juice and cooled chilli syrup. Push through a sieve to remove the seeds. In a large bowl, beat the ricotta and cream until very smooth. Fold in the guava purée, mixing well. Cover and chill for at least 2 hours. Churn and freeze in an ice cream maker. Once thickened, either serve right away, or store in the freezer to harden.

peach and amaretti ricotta ice

A deliciously smooth ice cream containing chunks of fruit and crushed amaretti biscuits for contrasting crunch.

Makes about 1 litre/1¾ pints/4 cups

5–6 good-quality ripe peaches
3 tbsp lemon juice
115 g/4 oz/generous ½ cup sugar
200 g/7 oz/generous ¾ cup ricotta cheese
150 ml/¼ pint/⅔ cup whipping cream
40 g/1½ oz amaretti biscuits, crumbled

Cut the peaches in half though the indentation and scoop out the stones. Slice into quarters and remove the skin with a sharp knife. Sprinkle with a little lemon juice as you work. Finely chop the flesh of two of the peaches, sprinkle with more lemon juice and set aside. Roughly chop the remaining peaches. Put into a food processor with the sugar and the rest of the lemon juice.

Purée until very smooth, then tip into a mixing bowl. Beat together the ricotta and whipping cream, then fold into the purée. Churn in an ice cream maker for 10–15 minutes or until the mixture begins to thicken. Mix the amaretti crumbs with the reserved chopped peaches, add to the mixture and continue churning. Once thickened, either serve right away, or store in the freezer to harden.

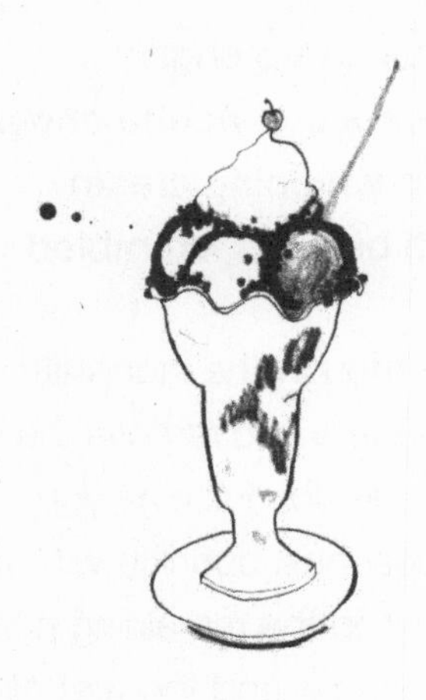

making ices with alcohol

- *Hints for Making Ices with Alcohol*
- *Ices with Alcohol Recipes*

hints for making ices with alcohol

- It's tempting to keep adding alcohol to the chilled mixture until you are sure you can taste it, but by then you will probably have added too much. A light touch is what's needed.
- You can always flood the plate with alcohol or pour some over the finished ice cream if you want a more intense flavour.
- Vodka has the least intrusive flavour, but a more assertive liqueur such as kirsch is better with stone fruits.
- Liqueurs such as Cointreau or Grand Marnier enhance custard-based ice creams or orange sorbets.

prune and armagnac ice cream

A decadently rich and alcoholic ice cream for a special occasion. Start at least 24 hours ahead of serving to allow time for soaking the prunes.

Makes about 750 ml/1¼ pints/3 cups

250 g/9 oz large pitted no-soak prunes
about 300 ml/½ pint/1½ cups Armagnac or brandy
2 tbsp caster sugar
150 ml/¼ pint/⅔ cup whipping cream

Put the prunes in a bowl with enough Armagnac to cover. Leave to soak for at least 24 hours or up to 1 week. Drain thoroughly, reserving the liquid, and chop the flesh very finely. Dissolve the sugar in 3 tablespoons of the drained Armagnac. (Use the rest as a sauce to pour over the ice cream.) Whip the cream lightly until the beaters leave a faint trail when lifted from the cream. Stir the prunes and the sweetened Armagnac into the cream. Churn and freeze in an ice cream maker. Once thickened, either serve right away, or store in the freezer to harden.

quince ice with poire william eau de vie

A rich, creamy, deep pink ice cream. The flavour of pears from the Eau de Vie is perfect with fragrant quinces.

Makes about 850 ml/1½ pints/3½ cups

600 g/1¼ lb quinces
500 ml/18 fl oz/2 cups Sugar Syrup (page 54)
150 ml/¼ pint/⅔ cup water
juice of 1 lemon
4 tbsp Poire William Eau de Vie
pinch of salt
1 quantity cold Vanilla Custard (page 50)

Quarter, core and peel the quinces. Slice the quarters into thin segments and put in a saucepan with the syrup and water.

Bring to the boil then simmer gently, covered, for 40–50 minutes until tender. Remove from the heat, drain the quinces and leave to cool. Put the quinces in a food processor and purée for 3–4 minutes until very smooth. Push through a nylon sieve to remove any residual grittiness. Stir in the lemon juice, eau de vie and a pinch of salt. Combine with the vanilla custard, then cover and chill for at least 2 hours. Churn and freeze in an ice cream maker. Once thickened, either serve right away, or store in the freezer to harden.

blackcurrant and cassis sorbet

Make the most of summer blackcurrants in this intensely flavoured garnet-red sorbet. It's fine to use frozen currants if fresh are unavailable.

Makes about 750 ml/1¼ pints/3 cups

450 g/1 lb blackcurrants (defrosted if frozen)
150 g/5 oz/¾ cup sugar
450 ml/¾ pint/1¾ cups water
2 tbsp cassis

Put the blackcurrants in a small saucepan with 3–4 tablespoons of water. Cover and simmer for 10 minutes until softened. Purée in a food processor, then push through a nylon sieve to remove the pips. Heat the sugar and water in a small saucepan, stirring until the sugar has dissolved. Boil for 5 minutes until the bubbles look syrupy. Remove from the heat and leave to cool. Stir in the blackcurrant purée and the cassis. Cover and chill for at least

2 hours. Churn and freeze in an ice cream maker. Once thickened, store in the freezer to harden.

pink grapefruit and gin sorbet

A glamorous pink sorbet with just enough gin to give it a kick.

Makes about 950 ml/1¾ pints/3¾ cups

3 pink grapefruit
500 ml/18 fl oz/2 cups Sugar Syrup (page 54)
juice of 1 lemon
2–3 tbsp gin

Thoroughly wash one of the grapefruit by scrubbing in warm soapy water. Rinse and dry. Finely grate the zest from the scrubbed grapefruit, taking care not to include any of the bitter white pith. Put the zest and sugar syrup into a saucepan and bring to the boil. Remove from the heat and leave to infuse until completely cold. Squeeze the juice from all three grapefruit –

there should be about 500 ml/18 fl oz/2 cups. Strain through a fine-meshed sieve. Stir the strained grapefruit juice, lemon juice and gin into the cold syrup. Cover and chill for at least 2 hours. Pour through a sieve to remove the grated zest. Churn and freeze in an ice cream maker. Once thickened, store in the freezer to harden.

cranberry and port sorbet

Smooth, intensely cold and very red, this is a stunning sorbet to end a festive winter meal.

Makes about 850 ml/1½ pints/3½ cups

250 g/9 oz cranberries
450 ml/¾ pint/1¾ cups water
225 g/8 oz/1 heaped cup caster sugar
5 tbsp ruby port

Simmer the cranberries and water in a small saucepan until soft. Pour the cranberries and their liquid into a blender and purée until smooth. Push the mixture through a nylon sieve and into a bowl. Add the sugar and port, and stir until the sugar has dissolved. Chill for at least 2 hours. Churn and freeze in an ice cream maker. Once thickened, store in the freezer to harden.

blood orange and cointreau sorbet

A refreshingly tart, flame-coloured sorbet. Serve by itself, or layer in tall glass goblets with whipped cream flavoured with finely grated orange zest.

Makes about 1 litre/1¾ pints/4 cups

12–14 blood oranges
2 tbsp Cointreau or other orange-flavoured liqueur
2 tbsp sugar, or to taste

Squeeze and strain the juice from the oranges. Add the liqueur and sugar, and stir until completely dissolved. Cover and chill for at least 2 hours. Churn and freeze in an ice cream maker. Once thickened, store in the freezer to harden.

making unusual ices

- *Hints for Making Ices with Herbs*
- *Ices with Herbs Recipes*
- *Savoury Ices Recipes*

hints for making ices with herbs

- The exquisite flavour of herbs is best reserved for water ices; the flavour is too delicate to compete with cream or eggs.
- Herb ices are only worth making with healthy, home-grown herbs that you have picked within an hour of starting to make the ice.
- To intensify the flavour, simmer the leaves in the standard sugar syrup and combine this with a decoction made by steeping the leaves in hot water.
- To appreciate these ices at their best, eat them on the day of making; they do not store well in the freezer.

angelica and rhubarb sorbet

This recipe needs fresh angelica, which grows like a giant celery plant. Make the sorbet in early summer while the angelica and rhubarb are still tender. Eat within 24 hours of making, when the flavour is at its best.

Makes about 750 ml/1¼ pints/3 cups

450 g/1 lb trimmed young rhubarb
100 g/3½ oz trimmed young angelica stems
600 ml/1 pint/2½ cups Sugar Syrup (page 54)
4 tbsp water
100 g/3½ oz/½ cup sugar
juice of ½ lemon

Slice the rhubarb and angelica stems into 2.5 cm (1 in) pieces. Put the angelica stems in a saucepan with the syrup. Bring to the boil, and simmer briskly for 5 minutes. Remove from the heat, pour into a bowl and leave until completely cold. Once cold, strain through a sieve to remove the angelica stems. Meanwhile, put the rhubarb in a saucepan with the water. Bring to the boil, then reduce the heat and simmer for 5 minutes until the rhubarb is tender. Remove from the heat and leave to cool a little. Purée the mixture in a food processor. Tip into a bowl and, when completely cool, stir in the lemon juice and the strained angelica syrup. Taste and add more sugar if necessary. Cover and chill for at least 2 hours. Churn and freeze in an ice cream maker. Once thickened, store in the freezer to harden.

rosemary sorbet

To appreciate the delicate herb flavour, eat the sorbet on the day of making or within 24 hours at the most. Use tender rosemary shoots from the tip of the branch rather than woody stems.

Makes about 1 litre/1¾ pints/4 cups

2 large handfuls of tender rosemary shoots
600 ml/1 pint/2½ cups Sugar Syrup (page 54)
500 ml/18 fl oz/2 cups boiling water
4 tbsp lemon juice
small handful of rosemary flowers (optional), green parts removed

Put half the rosemary shoots in a saucepan with the syrup and boil for 5 minutes. Remove from the heat, pour into a bowl and leave until completely cold. Once cold, strain through a sieve to remove the rosemary. Meanwhile, put the remaining rosemary in a teapot or jug. Pour in the boiling water. Immediately cover and leave to cool. Strain through a sieve. Combine the cold syrup, rosemary infusion and the lemon juice. Cover and chill for at least 2 hours. Churn in an ice cream maker for 15 minutes. If using rosemary flowers, add them now, and continue churning until thickened. Either serve right away, or store in the freezer for no more than 24 hours.

• variations

rose pepper geranium sorbet: Use 2 large handfuls of pepper geranium leaves. Add a generous grinding of black pepper to the syrup when boiling the leaves. Add the juice of 2 lemons to the final mixture.

lemon balm sorbet: Use 2 large handfuls of tender lemon balm leaves. Add the juice of 3 lemons to the final mixture.

lemon verbena sorbet: Use 2 large handfuls of tender lemon verbena shoots without any woody stems. Add the juice of 3 lemons to the final mixture.

quince and mint sorbet

Quince and mint are a heavenly combination. Use a strongly flavoured mint, such as spearmint or peppermint, and add it towards the end of churning.

Makes about 850 ml/1½ pints/3½ cups

600 g/1¼ lb quinces
600 ml/1 pint/2½ cups Sugar Syrup (page 54)
juice of 1 large lemon
3 tbsp chopped fresh mint

Quarter, core and peel the quinces. Slice the quarters into thin segments and put in a saucepan with the syrup. Bring to the boil, then simmer gently, covered, for 40–50 minutes until soft. Add a little water if necessary. Remove from the heat and allow to cool a little. Pour the mixture into a food processor and purée for 3–4 minutes until very smooth. Push through a nylon sieve to remove any residual grittiness. Stir in the lemon juice and leave until completely cold. Cover and chill for at least 2 hours. Churn in an ice cream maker for 15 minutes. Add the mint and continue churning until thickened. Either serve right away, or store in the freezer to harden.

savoury ices recipes

Savoury ices have a bite and sharpness that makes them particularly suitable for serving at the beginning of the meal to whet the appetite. Savoury water ices are also good for refreshing the palate between courses.

cucumber and lemon thyme ricotta ice

This is delicious served with a salad of strongly flavoured mixed leaves, such as rocket, red chicory and baby spinach. The contrast of texture and temperature is mouth-watering. Use within 24 hours of making, before the delicate flavours start to fade.

Makes about 850 ml/1½ pints/3½ cups

500 g/1 lb 2 oz tender-skinned cucumber, coarsely chopped
125 ml/4 fl oz/½ cup vegetable bouillon, such as Marigold
250 g/9 oz/1 cup ricotta cheese
2 tbsp lemon juice
2 tbsp chopped fresh lemon thyme
1 tsp sea salt
½ tsp freshly ground black pepper

Put all the ingredients in a food processor and process until smooth. Tip into a bowl, cover and chill for at least 2 hours. Churn and freeze in an ice cream maker. Once thickened, either serve right away, or store in the freezer to harden.

horseradish ice cream

Serve melon ball-sized scoops of this palate-tingling ice cream with a magnificent rib of roast beef. It's also excellent with smoked salmon or mackerel.

Makes about 600 ml/1 pint/2½ cups

250 ml/9 fl oz/1 cup whole milk
250 ml/9 fl oz/1 whipping cream
4–5 tbsp finely grated fresh horseradish
good pinch of salt
3 egg yolks, preferably organic
2 tbsp sugar

Heat the milk and cream in a saucepan over medium heat until steaming (80°C/175°F). Remove from the heat and add 4 tablespoons of the horseradish and a good pinch of salt. Stir, then taste and add more horseradish if you think it needs it. (Remember that freezing will dull the flavour.) Leave to infuse while you make the custard. Beat the egg yolks with the sugar for 5 minutes until very pale and creamy. Gradually add the warm horseradish mixture, beating well with each addition. Pour the mixture back into the pan and stir over medium–low heat for 4–5 minutes until thickened (82–83°C/180–185°F). Take care not to let the mixture boil. Cool the custard quickly by pouring into a bowl immersed in a larger bowl of ice cubes or very cold water. Once cold, cover and chill for at least 2 hours. Push through a fine nylon sieve to remove the horseradish shreds. Churn and freeze in an ice cream maker. Once thickened, either serve right away, or store in the freezer to harden.

Quick Tip

If you can't find fresh horseradish, use grated fresh horseradish from a jar.

celery and mascarpone ice

This delicately flavoured ice is best served on the day of making it, before the flavour fades. It's perfect with smoked salmon or a mixed leaf salad.

Makes about 600 ml/1 pint/2½ cups

2 heads celery, preferably organic
350 g/12 oz/1½ cups mascarpone
4 tbsp vegetable bouillon
juice of ½ lemon
1 tsp celery salt
¼ tsp freshly ground black pepper
few drops of Tabasco sauce

Trim the root and leaves from the celery and discard the tough outer stalks (use in soup or stock). Remove the strings using a swivel peeler. Slice the stalks lengthways into 2 or 3, then chop crossways into small dice. Purée in a food processor, then push through a fine-meshed sieve, pressing hard with the back of a wooden spoon to extract as much flesh as possible. You should end up with about 250 ml/9 fl oz/1 cup of sieved purée. Mix the purée with the remaining ingredients, then cover and chill for at least 2 hours. Churn and freeze in an ice cream maker for no more than 15 minutes – otherwise the mixture will turn to butter. Once thickened, either serve right away, or store in the freezer to harden.

bloody mary sorbet

For the best flavour use a good-quality tomato juice. You can also use a tomato-based vegetable juice.

Makes about 850 ml/1½ pints/3½ cups

750 ml/1¼ pints/3 cups good-quality tomato juice

6 tbsp Sugar Syrup (page 54)

juice of 3 limes, strained

2 tbsp balsamic vinegar

½ tsp celery salt

½ tsp freshly ground black pepper

a few drops Worcester sauce, to taste

a few drops Tabasco sauce, to taste

pinch of sea salt

Combine all the ingredients in a bowl or jug, mixing well. Cover and chill for at least 2 hours. Churn and freeze in an ice cream maker. Once thickened, store in the freezer to harden.

• variation

For a mildly alcoholic sorbet, add 2–3 tablespoons of vodka.

troubleshooting

- *FAQs*
- *Problems, Explanations and Solutions*

faqs

Q: My ice cream maker's instruction manual only gives measurements in cups. How do I measure dry ingredients this way?

A: Cook shops sell standard cup measures for measuring liquids and dry ingredients. Add the ingredient and level the surface without packing it down. Otherwise, for liquids, you can use a calibrated plastic measuring jug, allowing 250 ml/ 9 fl oz for 1 cup.

Q: I only want to make a small quantity of ice cream. How do I vary the ingredients and churning time?

A: If you want to make half the quantity, for example, reduce the ingredients accordingly. Because there will be less mixture to churn, the churning time will be shorter, but not necessarily halved. Be aware that manufacturers usually specify a minimum volume of mixture that can be successfully frozen. Check the instructions first.

Q: I am following a calorie-controlled diet. How can I make low-fat ice creams?

A: Thick Greek yogurt or ricotta cheese are only 10 per cent fat and can be used instead of whipping cream. You can also substitute skimmed or semi-skimmed milk for whole milk, but the texture will be less creamy, and the flavour not nearly as full.

Q: My ice cream doesn't seem to stay properly frozen in the freezer.

A: Use a freezer thermometer to check that your freezer is running at between -18° and -23°C. Store your ice cream in a proper freezer rather than the freezer unit in the fridge. The door is opened less often so the temperature remains stable.

Q: My mixture leaks out of the top of my machine when churning. What am I doing wrong?

A: You have filled the container more than three-quarters full, and the mixture has expanded during churning.

problems, explanations and solutions

Making ices is a precise process and common problems invariably arise. Fortunately, there is often a simple explanation.

The custard has curdled

- The mixture was overheated and the eggs began to scramble. Immediately strain the mixture into a cold bowl to break up the solidified particles.
- If the mixture is only just curdled, a quick whiz in the blender may help re-emulsify it.

The solid ingredients have sunk

- Ingredients such as chocolate, biscuits, nuts, dried fruit, or pieces of whole fresh fruit, should be chopped or broken into very small pieces and, ideally, chilled or pre-frozen before adding to the mixture.
- Don't add these ingredients until at least halfway through churning, after the mixture has firmed up. This keeps them from sinking to the bottom of the container, and prevents

crumbly biscuits from disintegrating completely.

The mixture is taking too long to thicken

- The mixture was not sufficiently chilled before churning.
- The built-in container or freezing canister was not chilled before starting to churn.
- The mixture contains too much sugar or alcohol.

The ice cream looks and tastes like butter

- Double cream was used instead of whipping cream.
- The cream was over-whipped before adding to the custard base.
- The mixture was over-churned.

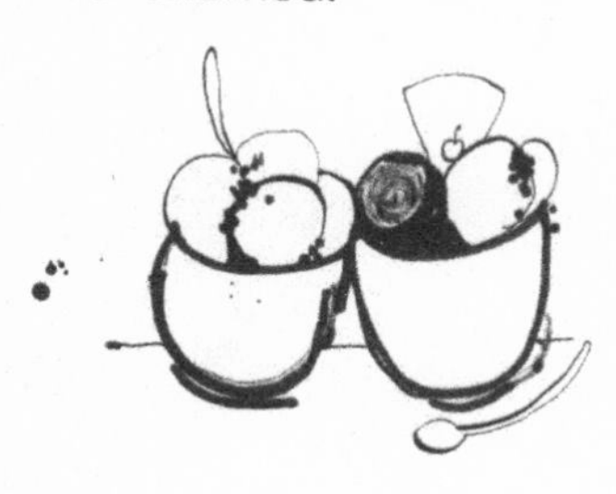

The ice cream is too hard and icy

- The mixture contains too much water and not enough sugar.
- It was not churned enough to break down the ice crystals.

The ice cream is gritty

- The mixture doesn't contain enough fat.
- The lactose in the milk has frozen.
- The custard base was curdled.

The ice cream has frozen unevenly

- The machine was turned on too soon before adding the base mixture, so the outer mixture has frozen too fast.
- The paddle is not scraping the sides of the canister properly.

glossary

Ageing

A term used in the ice cream industry meaning the length of time a pasteurised base mixture is kept before churning. Ageing improves texture, flavour, ease of whipping and resistance to melting.

Churning

Mechanically stirring the mixture during freezing to break down ice crystals and to add air.

Come to

The stage which ice cream should be allowed to reach so that it is just soft enough to scoop.

Granita

A coarse-textured Italian water ice that melts quickly. Best made by still-freezing since an ice cream maker would break down the characteristic large ice crystals.

Hardening

This refers to the further freezing of a partially frozen or soft scoop ice.

Over-run

A term used in the ice cream industry meaning the amount by which a mixture increases in volume from the chilled base to the finished ice cream. The increase is a result of introducing as much as 50 per cent air into the mixture.

Pasteurisation

An industrial process used to destroy harmful organisms by heating the mixture to a specific temperature for a specific time. The mixture is then cooled rapidly within a specific time and held at a low temperature until frozen. Home-made custards heated to 85°C/185°F and cooled rapidly are within safety limits.

Ripening

Another term for chilling the custard base, to allow the vanilla flavour to permeate fully.

Sherbet

Popular in America and similar to a water ice, but containing milk and/or cream as well as fruit.

Soft scoop

Ice cream that is soft enough to be scooped easily.

Sorbet

A French word for a water ice made with sugar syrup and flavourings – usually fruit juice or purée – but without egg yolks or dairy products. Sorbets traditionally contained egg white to stabilise or lighten the mixture but this is not necessary if you use an ice cream maker.

Still-freezing

A manual method of freezing in which the mixture is put in a shallow container in the freezer. The ice crystals which form around the edge as the mixture hardens need to be broken up periodically during the freezing process and mixed with the unfrozen liquid in the middle. This can be done with a fork or by putting the mixture in a food processor, processing until smooth then returning to the container for further freezing.